The Universal Frequency Theory

How the Energy of the Universe Flows in a Closed Loop

Kenneth W. Liskey Sr.

The Universal Frequency Theory

Original Publication Date: 09/08/20

A Case for Energy being able to Efficiently pass over itself in All Directions Throughout Space.

The ***Universal Frequency Theory*** dictates that all of the universe's energy is contained within, while remaining in a constant state of expansion and contraction where both occur simultaneously. This frequency, of expanding dark matter energy in the physical realm, is counterbalanced by the contraction of light matter energy in the spiritual realm and can be described by the following equation: ***Dark matter energy expansion = Light matter energy contraction***.

As such, the above dynamic comparison represents the constant motion of universal energy within a closed loop system and implies that an energy imbalance exists between the dark and light matter energies at any given point in the universe. In actuality, this energy disparity between the physical and spiritual realm is what facilitates continuous energy movement from one state to the other, as the uniflow frequency attempts to reach equilibrium.

To further build our understanding of this equation, one must consider the idea of dark matter energy in the physical realm, as spiraling out from the center of the universe in all directions. Yet, dark matter energy remains tightly organized in threads, consisting of end-to-end vortexes.

These in-line vortex connections occurring within each thread, are ionically interactive due to the influence of the coinciding dark matter particles, as they work to bind the dark matter energy into an interwoven fabric. All the while, this fabric is collectively propelled towards the outer reaches of our universe in ordered waves of Fibonacci spirals.

In theory, the dark matter energy flows over a group of rotating dark matter particles, before converging and arcing across towards the next grouping, to create each end-to-end in-line vortex chain within a thread. These newly forming energy threads remain super compressed by the strong opposing forces from outside of the universe, causing the threads themselves to organize in woven bundles as they twist their way towards the universe's perimeter.

Still, each closely stacked end-to-end energy vortex thread is progressively expanding in all directions, to occupy the available space. Effectively, when these interwoven threads reach the edge of the universe, their ionic connections are now stretched to their limit. Of equal importance is the negative space around each thread, which expands in the same proportionate way.

Eventually, the outer wave of these intertwined threads encounters the shifting perimeter of the universe, ebbing and flowing like an ocean tide. The variable nature of the edge transpires because our expanding universe, interacts with the oppositely expanding forces from the surrounding universes, which completely engulf our universe. Here the dark matter threads behave much like water molecules along the banks of a river, as the first wave produces a buffer zone between the independent universes.

The subsequent dark matter waves to follow will passively compress this buffer zone and force abrasion against the fabric of the surrounding universes, tearing the over-stressed fabrics into virtually massless, individual nano-particles and releasing an immense amount of universal energy. These supercharged nano-particles are then centripetally forced back towards the center of the universe as light matter energy.

This sudden implosion of light matter energy is in part due to the oppositely expansive pressure from the surrounding universes, coupled with the remaining ionic pull, (or elasticity), from the corresponding dark matter threads. Once the outward momentum swings inward, the contracting light matter is drawn back to the bundled dark matter threads, where the nano-particles rotate around the exterior of these threads, similar to a planet orbiting around a star, as the light matter energy spirals inwards towards center of the universe.

In particular, light matter energy is utilizing the existing propulsion action of the dark matter energy, but in reverse. Therefore, light matter propulsion is the reaction to the action and spirals in the opposite direction of the dark matter end-to-end vortex threads, as an implosion. Light matter energy can reach speeds exponentially beyond the speed of light, due to the negligible friction loss of these infinitely smaller energy particles, as they rotate around the dark matter fabric in a vortex.

Once the light matter energy implodes to the center of the universe the nano-particles condense and reform as dark matter energy, where they cohere to new dark mater particles and reestablish their in-line vortex threads. In this way, the process is energized to start over again. Essentially, the dark matter energy spirals outward utilizing end-to-end vortex chains, while the light matter energy simultaneously contracts inward using these same vortex forces, in a perfectly choreographed but opposite motion.

The concept is similar to alternating current, except in this instance, the energy moves over itself in both directions, rather than through insulated wires. In effect, the dark matter energy's overriding attraction towards the coinciding dark matter particles is so strong, the resulting bond acts as an insulator to separate the comingling energies.

Yet, this energy transformation process may create the gases, light, heat and microwave energy found in space, as these tightly juxtaposing energies scrape against each other. Though technically, there are no direct collisions of molecules because super rotating nano-particles within an energy vortex will deflect the impact, making this the most efficient way for oppositely flowing energies to coexist.

Also, the friction taking place from this energy transformation vortex may generate the matter we touch, so our physical world might merely be a byproduct of the expanding dark matter abrading against the imploding light matter, as they maintain a universal equilibrium. In universal terms, matter, light, gas, heat and microwave energy represent no net gain or loss in the universal energy balance, but instead are a by-product fusion of dark and light matter. This energy progression where the imploding light matter energy reforms as expanding dark matter energy, is what releases a seemingly never-ending source of universal motion.

Therefore, the universal frequency is on a perpetual loop and can be characterized by the equation: ***Dark matter energy expansion = Light matter energy contraction***, where = represents the inversely proportionate balance between the two, and must be measured with a sliding scale in relationship to location within the universe. Only then can we adequately capture the dynamic nature of this frequency.

Since matter, light, gas, heat and microwave energy are all hybrid energies resulting from the fusion of light matter and dark matter, they are not pure energy and hence can be temporarily excluded from the universal energy equations. Primarily, any friction loss happening between these oppositely flowing energies can be viewed as being shared equally and so, mathematically cancels itself out. Given that the maximum speed of dark matter energy expansion is capitated to the speed of light, or ***C***, we can represent the maximum distance traveled by dark matter energy during a specific time period as the speed of light.

Furthermore, all things in the physical realm contain mass and dark matter energy is no exception, so then ***C ÷ m = Dark matter energy expansion***. The mass of the dark matter fabric or ***m*** must divide ***C***, to factor in the ionic drag occurring as energy moves through space in the physical realm.

This basic equation, accounts for the perpetual stretching of the dark matter fabric out to the edge, where lower densities of dark matter will reside at any given point of the universe, as measured from the center. Less overall mass would imply lower ionic resistance, causing dark matter velocity acceleration.

On the other side of the equation, light matter energy must contract minimally at the speed of light squared, in order to counter-balance the dark matter fabric expansion. Since light matter energy moves minimally at the speed of light squared in the spiritual realm, time would be unperceivable, though we can still quantify the action of these nano-particles surrounding the dark matter threads, as they return towards the center of the universe.

We can represent this rotation, with the coefficient $\boldsymbol{\pi}$. Pie then signifies the light matter energy rotating around a dark matter thread, as it establishes a corresponding light matter thread when contracting to the center of the universe. Lastly, light matter energy mass in universal terms, embodies the smallest unit possible, so we can represent light matter mass with the number ***1***.

In other words, light matter energy collectively moves with no appreciable friction loss. Subsequently, the minimum velocity of light matter contraction equals the ***(speed of light*** ^{2} ***x*** π***) divided by 1***.

Therefore, at the universe's edge, the minimum velocity of light matter energy, or $\boldsymbol{C^2 x \pi}$ coexists beside the maximum velocity of dark matter energy, or $\boldsymbol{C \div m}$. So, the equation $\boldsymbol{(C^2 x \pi) x m = C}$ denotes the differential between the velocities of light matter energy and dark matter energy, as expressed at the edge of the universe, where the symbol = proportionately represents the necessary imbalance to generate and sustain a frequency.

If we can utilize dark matter density and velocity to determine its energy, we can also extrapolate light matter's potential for possessing far greater amounts of energy than dark matter at the universe's edge, as evidenced by the above-described Universal Differential equation.

Consequently, when calculating dark matter energy at a specific point in the universe, the equation could be represented in the following manner: $[(C^2 \times \pi) \times m]^k \div C =$ ***UDME or Universal Dark Matter Energy***. The power of ***k*** represents the distance measured from the edge of the universe, and increases proportionately when moving towards the universe's center in a straight line, with ***k*** represented as the power of ***1*** at the universe's edge.

The coefficient ***k***, factors in the exponentially increasing force, required to transport energy through a vortex. By comparing specific points of this equation from the edge to the center, one can project that the center of our universe will contain the bulk of dark matter energy and mass, in their most condensed forms.

Similarly, when calculating light matter energy at a specific point in the universe, the equation could be represented in the following manner: $\{[(C^2 \times \pi) \times m]^k \div C\}^w$ = ***ULME or Universal Light Matter Energy***.

Likewise, the power of ***w*** represents the distance measured from the edge of the universe, and increases proportionately when moving towards the center, with ***w*** denoted as the power of ***1*** at the universe's edge. The coefficient ***w***, factors in the exponentially increasing force, required to transport energy through a second vortex.

Contemplating the idea of this universal frequency as a uniflow closed loop operating with opposite poles, a second vortex coefficient is needed to accurately depict this two-beat frequency, where energy is released at both ends. Along this line of reasoning, the center of the universe contains the bulk of light matter energy in its most condensed form and drives the dark matter fabric expansion, while the energy transition at the universe's edge, generates a second pole.

In theory, the corresponding dark matter and light matter threads, remain super-compressed against each other due to pressure from the surrounding universes, at least until the waves approach the outer reaches. This density variation of energies from the center of the universe to the edge in a progressive manner, is reflected in the Universal Dark and Light Matter Energy equations.

Yet, it is the close juxtaposition of these stacked threads throughout a large portion of the universe, which makes the organization of like-particles possible, within the physical realm. We can readily observe this universal phenomenon, where like-particles such as light, are gathered together and then transported over great distances in waves.

Such organizational predisposition within the universe is possible, because the in-line dark matter vortex threads can push forward, pull backwards, and seamlessly combine in any direction. Also, new threads may spin off from the existing threads to fill the expanding available space. This infinite versatility allows like-threads to attract and connect ionically in all directions, which automatically draws like-particles together.

Once like-particles are congregated, the particles themselves have the capability to form ionic connections of their own, making it possible for the vortex chains within the threads to stream these bonded like-particles in what we term a wave frequency.

In some instances, where individual like-particles remain too far apart to make an ionic connection, they may still connect in wave form by utilizing either the same vortex thread, or an adjacent synchronized thread. This would explain the vibrational synchronicity of particles seemingly disconnected by distance.

However, even though dark matter energy is the prime mover of the physical realm, we cannot perceive it. Nonetheless, the fact that we can see, hear and smell individual frequencies could be viewed as hard evidence of these infinitely-directional dark matter threads. Particle transport within these vortex threads can reach, but most likely not exceed the speed of light within the physical realm.

Conversely, when light matter energy implodes around the established dark matter threads on its way to the center of the universe, the first layer will abrade against the dark matter fabric. This micro-friction zone, equally involving both energies, allows the vast majority of light matter energy to slip-flow through the negative thread in a friction free manner and return to the center.

Essentially, all negative space existing around the dark matter threads can be utilized by the light matter energy, providing a path of least resistance back towards the center of the universe. A point worth noting is, when the adjacent dark matter threads synchronize together, so does the negative space around them, aiding in negative thread propulsion.

In other words, the synchronization of negative space dramatically increases the implosion capabilities of the light matter threads because the entire negative space now consists of end-to-end vortexes, capable of pulling back towards the center. If so, this synchronization process would be a critical organizational and transport mechanism in the spiritual realm, as well as in the physical realm.

Interestingly, the micro-friction zone interaction between the co-mingling light and dark matter threads, might actually further energize the adjacent dark matter fabric. Large scale evidence of such energy exchange possibilities is the supermassive black holes, imploding at the center of each galaxy.

Here huge concentrations of light matter streams spiraling towards the center of the universe, may transfer some of their energy to fuel the all-consuming dark matter vortex, as the opposite forces ride over each other in a synchronistic way.

Simultaneously, this process might fuse pure energy to form the gas, light, heat, matter and microwave energy of each galaxy. Given enough time, supermassive black holes will recycle all fused energy back into light matter, with little net effect on the energy equations.

Similarly, stars at least in part, can self-recycle once the gravity vortex created from a collapsing star condenses the corresponding light matter energy, to a point where the black hole shifts to the all-consuming negative gravity phase.

Such negative gravity may occur, when a light matter implosion supersedes the dark matter vortex and this energy shift will ultimately force the recycled hybrid energy back to the center of the universe. This process seems to be mimicked by all vortexes.

Without such interaction, vortices could not form in the physical realm including hurricanes and tornados, though these smaller vortex formations will most likely not establish enough negative gravity to recycle hybrid energy. Instead, air and water vortexes seemingly exist to allow opposing masses to seek equilibrium.

Subsequently, the frequency of universal energy is one of expansion and contraction and consists of pure perpetual motion, fueled by the constant injection of light matter energy throughout the progression.

Whereas the movement of the physical components within the universe, operate under a time dependent process and are eventually recycled. In the case of planetary rotation, the revolving mass will fling the adjacent dark matter adhering to the planet's surface out away from the center.

At the same time, universal dark matter energy expansion is applying constant pressure from the center of the universe and instantaneously fills this void, but in doing so abrades against the fleeing dark matter. Hypothetically, the overriding action between these two opposing dark matter forces fractures the dark matter matrix, causing a short circuit to spark a steady stream of light matter energy into the physical realm.

This light matter arc, occurring between the opposing energy flows, is contained by the expanding dark matter pressure, which instantly squeezes the light matter back into the fractured thread. In this way, the overriding of these contrasting dark matter forces is what initiates planet rotation, along with the bending of space around the planet, while the light matter arc injects energy to perpetuate the motion.

Likewise, our rotating sun flings dark matter away from its surface on a tangent from the center and such phenomenon might explain the propensity for like-particle sun rays to travel in a straight line to the Earth.

This action where dark matter is flung away from the center of a rotating mass is what we understand as centrifugal force and the immediate filling of this void we know as gravity, (or centripetal force), yet the dark matter energy is merely seeking equilibrium. This harmonic overriding of these oppositely moving dark matter forces, releases light matter to continuously energize planet, star and galaxy rotation.

Accordingly, centrifugal force, gravity and rotation are inter-dependent, and operate collectively as a classic case of cause and effect. Essentially, this fracturing or short-circuiting of energy will remain in place until the planet is recycled, in most cases via a corresponding supermassive black hole, which could be considered the ultimate entropy.

As for practical application, this short-circuiting of energy may be replicated to a lesser degree, by abrading adjacent air, water or microwave molecules using centrifugal and centripetal forces. When these opposite forces are placed against each other, the rotating molecules scraping counter to one another, may be capable of coaxing the surrounding dark matter energy into releasing some of this above-described universal energy.

Mechanical inventions would most likely not be able to short-circuit these dark matter threads in the same manner as planets, but rather would release smaller amounts of the stored energy within the dark matter particle itself. This trickle release of dark matter energy, would be instantaneously reenergized by the commingling light matter thread. So, the efficiency of each individual procedure will determine how much universal energy can be released and for how long the spark will continue, before more external energy must be supplemented to the procedure.

Developing a working model of the universe, along with balanced equations to describe universal motion, came from understanding the existence of opposite poles, where energy moves back and forth between the spirit realm and the physical realm in a closed loop.

Broadly speaking, the Universal Frequency Theory contradicts the big bang theory, for it predicts that a closed loop universe was formed, when the convergence of the surrounding universes commingled their light matter energy to form a separate entity.

Afterwards, the dark matter and dark matter energy of our new universe spiraled out collectively as a fabric to reach their maximum sphere of influence, redefining the sphere of influence for each of the surrounding universes.

The resulting closed loop system is what feeds a constant supply of light matter energy to the center of our universe and sustains the expansion of the physical realm. In this way, the universe operates with high efficiency, similar to alternating current, rather than the less efficient direct current model, suggested by the Big Bang explosion.

In consequence, this dynamic scenario, where equilibrium is sought from imbalance, provides an underlying framework for energy to remain in continuous motion, without decay. Such perpetual movement enables the soul, consisting of light matter energy, to transform from life to afterlife and back to life again, indefinitely.

This constant energy conversion may explain why all souls are seemingly subjected to the law of chaos pervading the expansive physical realm, along with the law of reorder permeating the imploding spiritual realm, otherwise referred to as Yin and Yang.

Considering the close juxtaposition of these opposing energies one could infer the endless potential for temporary cross-overs between realms in the form of spirit entities, dreams, visions and E.S.P. Unfortunately, in the conscious state we can only witness such phenomenon under altered conditions because our brain network is generally not designed to perceive energy beyond the speed of light.

The limitation is in part due to thought being a physical act, comprised of particles, chemical reactions and connections, all capitated to the speed of light. This series of action, renders the common thought process far too slow to perceive the spiritual realm. Fundamentally, when thought generates particle motion in the physical realm, the action is subjected to the universal law of attraction.

In essence, dark and light matter energy will indiscriminately react in combination to the thought process, by first organizing our individual thoughts together and then working to juxtapose our thoughts with the like-thoughts of others. Theoretically, like-thought synchronization, would eventually draw like-souls together within the physical realm and this progression of soul connection, might explain the seemingly randomness of attraction towards other people.

Unique to the physical realm, congregated like-souls actually have a choice to form an ionic bond between themselves, a feat which is doubtful in the spiritual realm because energy is moving too fast. Such physical cohesiveness between individual souls, then makes possible the streaming of soulmates and soul packs from life to afterlife, in a synchronized energy wave. In principal, this process is identical to the gathering and streaming of other like-particles, which as described, is a phenomenon we can readily observe in the physical realm.

These synchronized soul energies would then share an ionic overlapping of frequency, possibly explaining the otherwise random comings and goings of souls in the physical realm, where one like-soul moving from life to afterlife, or vice versa, would eventually facilitate another to follow.

What confuses most of us is the lack of time in the spirit realm, causing us to assume there is no connection between individual deaths happening many years apart, when in reality, a death occurring decades later would appear instantaneous from the other side. As such, our choice of thought content may dictate what we learn, who we meet, and in due course how our soul evolves over countless lives and afterlives.

Effectively, all thoughts and actions in the physical realm will impact the dark matter matrix. The imprint is instantly recorded in the light matter, since light matter energy conforms to a corresponding dark matter thread, as it returns to the center of the universe.

This stardust recording of our soul's ever-changing light matter frequency, is continuously streamed as a subconscious energy every second of our existence, yet we cannot perceive this perpetually moving knowledge in our conscious state.

Quantified differently, in this life we are consciously hard-wired to receive our own individual subconscious frequency, which is an energy collection of our thoughts and actions, compiled from our lives, both past and present. Counterintuitively, our memory physically records the thoughts and actions of this lifetime as well, at least in part, allowing us conscious state access to our current life experiences.

Notwithstanding, even our most recent experiences in this life, are immediately recorded in the light matter energy, and then added to our subconscious frequency. Just as the rotating action of a planet interacts with the surrounding dark and light matter threads, our thoughts and actions have the same effect and the recording of such events will exist in the light matter energy, as long as our soul remains in motion.

So, one can imagine the core of our soul is the combined knowledge of our experiences congregated together and during life, collectively transported from the center of the universe out to the physical body.

While during the afterlife, this same soul energy is perpetually streamed from the center of the universe out to the edge, always moving within a closed loop. This steady stream of bundled energetic light matter threads comprises our subconscious soul frequency and, constitutes our lifeforce.

Such lifeforce energy transcends both realms and is the underlying spark igniting the physical act of internal thought. Devoid of this energy, the brain cannot fully function and becomes vegetative, because the brain must interact with this subconscious energy to generate complete thoughts. If so, a healthy being may possess the ability to use this lifeforce energy to project thought outward, through meditation or the third eye.

The notion of projecting thought is found in many ancient teachings and was evidenced through scientific observation, where simply viewing particles in motion changed the outcome.

A real-world application of this rudimentary telepathy is employed by futuristic robotic limbs, controlled wirelessly by thought. In theory, since internal thought is encumbered by brain synapses, the electrical thought field must be manipulated to expand beyond the body in order to expedite the thought process.

Given that the brain's electrical thought field is what most likely contains our subconscious soul frequency, when we project thought outwards, our soul frequency extends with it.

Unintentional examples of such an occurrence are outer body experiences arising from either a near death circumstance or exposure to a hallucinogen, where the connection between the soul frequency and the brain becomes partially interrupted.

During these events, the being feels out of body because the brain no longer contains the entire subconscious soul. Most likely, the brain is receiving just enough of the soul's frequency to ionically capture the remaining energy in the extended electrical field surrounding the body and, consciousness transpires only after the reception of this energy is once again corralled to reside fully in the brain.

To reiterate, our soul frequency or lifeforce exists as subconscious energy and fires the brain to create consciousness, whereby the brain is in part a mechanical receiver of this impulse and cannot fully function independently.

Clairvoyant individuals may temporarily circumvent the carnal process of internal thought with this hybrid method, where they are able to concentrate their lifeforce beyond the internal parameters, onto those around them.

One can envision, if the mechanics of the brain are temporarily circumvented, the light matter energy of the soul could actually project thought, and such an accelerated process from this altered consciousness would enable the individual to connect energetically with other living souls.

Spirit energies, which are perpetually streaming at high-speed intervals far too fast to perceive in the conscious state, would also be partially accessible.

But at such high frequencies, the spirit realm is out of sync with the physical realm and so contact with spirits remains fleeting. In practice, large-scale co-ordination of projected thought could facilitate the ionic overlapping of all soul frequencies and this process may eventually synchronize every being into the physical realm together, all at the same time.

The implication from this massive arrangement of overlying soul frequencies is; the sum total of their knowledge will be combined, at least on a subconscious level. So then, the act of choice in the physical realm either overlaps the soul's frequency with others, where an ionic bond can form, enabling multiple souls to stream as one or, the soul can choose aloneness.

Physical realm synchronization may be a critical component to the success of the soul's evolution, for the spiritual realm is timeless and a second is no different than 10,000 years, making a chance encounter with a known soul or any soul for that matter, highly improbable. Hence, we must ionically connect in the physical realm, in order to form lasting soulmates.

Yet, the paradox is, the more we connect and synchronize, the greater the populations accrue in the physical realm. Regrettably, explosive population growth is inversely proportionate to the sustainability of life in a two-pole universe. On the other hand, a complete synchronization of all soul frequencies through mass-projected thought could potentially congregate all past and current knowledge.

Some of this information may then be accessible in this altered, collective conscious state because such a shared thought stream, extending beyond the confines of the brain, will exceed the speed of light. In this cooperative way, we may be able to receive clearer glimpses of the ever-streaming spiritual realm information, surrounding us.

Creating this wireless connection between souls would require increased bandwidth for reception and transmission, which may explain why the human brain currently has so much unused brain matter.

Shared, projected thinking could in time grant us the use of these dormant portions of our brains, leading to the evolvement of telepathic communication and could at some point provide us with the ability to solve all of our societal problems.

Of equal importance, collective thoughts projected outside of the body vessel, would concentrate a great deal of light matter energy within the physical realm. There is the potential for this concentration from all intelligent life forms to short-circuit the adjacent dark matter matrix, forming a fracture similar to a rotating planet, but on a far greater scale.

Theoretically, during thought projection, the soul frequency no longer returns to the center of the universe from within the body. But rather, the energy extends out through the dark matter fabric in a thought stream, only to return to the universe's center after connecting with the other intended souls, even if those souls exist on the opposite side of the cosmos.

When we consider this feat from a mathematical perspective, this theory predicts that light matter energy travels from the body back to the universe's center at: $\{[(C^2 \times \pi) \times m]^k \div C\}^w$. To extend thought through the dark matter matrix, the formula $\{[(C^2 \times \pi) \times m]^k \div C]\}^w \times l$, would describe this act where l is represented as 1 at the physical body and grows proportionately with the distance a thought is projected beyond the body.

If so, a collective thought on a universal magnitude is quite possible, due to the speed with which light matter energy can travel. Such a collective universal thought could instantaneously concentrate immense amounts of light matter, pinpointing an energy vortex in the opposite direction.

This light matter vortex could develop high orders of gravitational pull and might ultimately reverse the universal poles. In another manner of speaking, a universal collective thought vortex could generate a supermassive black hole in reverse at the point of convergence, causing the physical realm to be flooded with light matter energy.

Once this reversal occurs, light matter energy would perpetually infuse all things in the physical realm, and possibly allow souls to exist in both realms, simultaneously.

A universe evolving to a third and vastly advanced state, will have no bounds, no physical restrictions and no time.

This concept, predicting a massive procession of light matter, validates man's seemingly inherent belief that Nirvana awaits the human race. Yet, this idea produces a conundrum where the population of the physical realm must reach a critical mass, before a synchronized thought frequency from all intelligent life can bring about the bypassing of our current two-pole system.

Not surprisingly, the outcome of the universe would ultimately be reduced to free will. For indeed, if every soul who has ever lived must simultaneously inhabit the physical realm in order to enact this universal shift, then collaboration is the only way intelligent life avoids extinction from our ever-growing populations. Perhaps this misunderstood conflict over collectivism versus individualism, is the true struggle between good and evil within the physical realm.

On a separate note, alien beings may very well be capable of reaching the Earth by utilizing the centrifugal force found in the outer bands of a nearby black hole, to slingshot their craft across the universe. Though hypothetically, according to this theory, it is not possible to exploit the implosion of a black hole as a wormhole for travel in the physical realm, because particle transport above the speed of light is the very definition of the afterlife.

Therefore, such an option is only available in the multi-dimensional spiritual realm, where each dimension is distinguished by its individual frequency or energy velocity, and may be accessible through these negative gravity implosions.

Nor can one reverse time, since all things physical must expand with the dark matter fabric, including bent space. Antidotal evidence to support this notion of universal energy expansion as dictating the present moment, is the current scientific observation of matter being driven to expand at a growing rate, by this very energy.

We can sense this phenomenon, whereas youngsters we perceive a year to be an eternity, while as adults we will view this exact same number of days to be much faster. Invariably, dark matter energy expansion acceleration is experienced innately within each lifetime.

Similarly, the surrounding space impacted by the rotation of a black hole, will not affect time, for even warped space in the physical realm is subjected to the higher universal law of energy expansion. And this multi-directional, spiraling dark matter fabric expansion into the ever-present moment is pure time.

In actuality, our calendar cycle, which is dependent on the physical properties of our solar system, is really a space-time relationship measurement comparing various planet and star rotations to universal energy expansion and, is a separate metering of time. Thus, another planet existing under completely different day and year cycles from those of the Earth, will mark time differently, but will remain in sync with our present moment, there is no escaping it.

Equally obtuse is the notion of man's ability to outrun universal expansion into the future, for matter transport cannot exceed the speed of light in the physical realm. Basically, the whole concept of linear trajectory enabling one to time travel is a human construct where we arrogantly believe, if we could travel backwards or forwards in time, we could orchestrate a better outcome. When in reality, improvement can only ensue with the advent of time and this gift of the ever-present moment is one where we make mistakes, die and then live again, until eventually, we evolve.

Under this theory, the idea of linear travel having an effect on time is erroneous because although such a concept may reflect matter's centrifugal expansion outward, as resulting from the rotating universal energy, it does not take into consideration the dark matter energy's multi-directional expansion.

In effect, dark matter's entanglement with the coinciding dark matter energy enables this collective fabric to react differently to universal rotation, by allowing the fabric to build up in super-compressed bundles of interwoven threads, spiraling away from the center in all directions.

One can easily imagine space travel from the universe's edge towards the center, as seemingly moving against the outward expansion of matter. However, a problem arises when we consider the idea of this energy fabric as spiraling outwards in all directions simultaneously, where we are only traveling in a single direction, with no net effect on time.

To further the argument against time travel, if we presuppose our super rotating universe to be generally in the shape of a sphere and, we can guesstimate a distance from the edge to the center, then our universal energy equations can approximate the energy residing at any given point, as measured along a ray.

For light matter particles, which are the smallest particles in the universe with their mass represented in the equation as 1, we can take the average between the edge and center to approximate the energy of these nano-particles.

Next, since the formula for the volume of a sphere is $V = 4/3\, \pi\, R^3$, we can apply our dark and light matter calculations to the universe's volume to determine the approximate dark and light matter energy of the universe.

Such massive numbers should clarify the impossibility of overcoming these powerful, multi-directional vortex forces. Hopefully, once humans accept their technological inability to physically access the past or the future, we will learn to better focus on what is right in front of us, the present moment. As a side bar, once we determine the total amount of hybrid energies existing in the physical realm, we can then add them to the total amount of pure energy.

This will permit us to approximate the sum total of energy in the universe. Again, this theory stresses the importance of segregating pure energy from hybrid energy when calculating the universal frequency.

While pondering the mysteriousness of universal energy, one may still question why dark matter energy expansion accelerates out to the edge, when undeniably the vast amount of dark matter resides at the center of our universe.

To answer this query fully, first and foremost, we must consider that during the formation of our universe, the light matter energy convergence of the surrounding universes, reached equilibrium by establishing super-rotation at our universal core.

This super-rotation of the core is now self-perpetuated by the continuing implosion of light matter energy, overriding the expanding dark matter energy, as previously described. Such a congregation and transformation of rotating energy, implies a velocity of zero at the point of convergence as found within the universe's center.

A simplistic analogy from our perspective on the Earth, would be like viewing a rotating chorus line of ice skaters, where each skater out from the center, must circle the ice faster to keep up with the line. Except, cosmically speaking, it is our universal energy fabric, spiraling out in Fibonacci tangents away from the center simultaneously in all directions, which is forced to accelerate in ever larger rotations out to the edge.

The astrological physics of this phenomenon, coupled with the dark matter's lowered ionic friction due to the ever-diminishing dark matter mass as it stretches out to the edge, combines to promote the fabric's acceleration.

Though we cannot overcome this multi-directional dark matter fabric expansion, we can most certainly manipulate it. Namely, the above referenced bent space should be exploitable and portals may be available for use as short cuts during space travel.

Additionally, the cause and effect associated with the interactions between gravity, centrifugal force and rotation are malleable, and we should be able to mechanically bend space directly adjacent the ship to reduce travel distance.

We may also be able to manipulate the dark matter surrounding the spaceship, to lower the ionic resistance during space travel for efficiency, along with exploiting dark matter energy for propulsion to make long distance space travel possible and even practical.

No doubt, since the odds are overwhelmingly in favor of extraterrestrials existing in the universe, some species have already mastered these transport mechanics.

Unquestionably, we should embrace the idea of alien visitations on the planet Earth and at least consider the notion of our race descending from alien species, whom have most likely developed an evolved state of consciousness. This crossover connection could be critical to our survival, for in view of the Earth's rapidly deteriorating conditions, man may not be able to develop quickly enough to solve his own problems and may need to rely on his alien brethren for other-worldly solutions.

Living as distant relatives of such advanced creatures, one would expect some physical similarities between us, although their evolving alien genetics would diverge from ours, in part due to the length of their species existence, circadian cycle and planet location within the universe.

But, even if alien bodies are designed to live longer in order to operate on unique soul frequencies, aliens are still subjected to the same universal law of expansion into the present moment, for it is absolute. And aliens will live and die and live again as we do, albeit on separate frequencies, contained by unique vessels.

Obviously, we are physically independent creatures, yet any genetic differences between species should not obscure the idea of alien soul's containing the same light matter energy, as we do. So, for all practical purposes, we are one in spiritual energy. This commonality should be more than enough to unify enlightened beings, with the understanding that each soul must be an active participant, if the Nirvana equation is ever to reach its full potential of removing the physical limitations of this realm.

Furthermore, since the light matter energy of our universe was born from the convergence of other universes, we too are spiritually connected to the species of the surrounding universes.

Boding the question of how far-reaching soul synchronization must extend, before our collective vibration resonates to the frequency of the Higher Power and aligns us with this ultimate energy source. Perhaps, we have known all along the definitive and far-reaching truth behind the words, "love thy neighbor, the same as self."

The terms vibration and frequency are often bantered about with no real understanding of their definitions. Under the rudimentary terms of this theory, the subconscious, which is a light matter energy recording of our existence, streams in a continuous frequency loop from the center of the universe into our brain.

As this energy frequency passes through our body's electrical field, a vibration is generated whereby both the frequency and the vibration are uniquely distinguishable to the individual.

Fundamentally, all matter is in vibration, suggesting that everything contains varying degrees of subconscious energy. These vibrations impact the surrounding dark matter matrix, allowing this omnipresent fabric to react, by synchronizing and organizing all likeness together.

Indeed, this like-vibrational phenomenon may actually determine the elemental makeup of all matter in our physical world and could be responsible for regulating matter's ionic capacity to distinguish itself into recognizable forms.

To comprehend how vibration could affect particle physics under this theory, we must first consider quark particles, which are the building blocks of protons and neutrons. Initially, the quarks must organize together to form these two main components found within the nucleus of an atom.

If we can assume that the light matter energy within like-quarks travels at the same speed, then a vibrational synchronicity will be established between them, allowing dark matter to congregate these like-quarks into close proximity.

In musical terms, these like-quarks would vibrate with the same notes. By adding dissimilar notes to this vibration, the congregating quarks could be altered slightly, creating micro-density disparities in order to differentiate an up quark from a down quark, etc. Interestingly, this theory would dispel the notion of gluon quarks, by assuming that the all-pervasive pressurized dark matter fabric is what actually pushes the like-vibrating quarks together.

Investigating the atom's makeup further, this theory predicts the possibility of protons and neutrons both starting out with an equal mass, since protons are only fractionally less in mass than neutrons. However, once the protons and neutrons are congregated together at the atom's center, the abrading action from their independent rotation could marginally reduce the proton mass and produce micro-particles as a byproduct.

This might occur when a point on the free spinning neutron contacts the surface of the adjacent free spinning proton, and gouges out a negative imprint of itself. Depending upon the axial rotations of both the neutron and the adjacent proton, along with the positioning of the point and gouge in relationship to these axial rotations, a gear-like scenario could emerge where the two independent rotations would syncopate the point with the gouge.

Such a configuration would have a great number of combinations which would require a minimum of one rotation, but could also necessitate more than one rotation, in order to complete the syncopation process of the nucleus. Pragmatically speaking, this 3-dimensional gear-like arrangement would also help to stabilize the nucleus, by physically realigning the individual components to augment their energetic attraction.

The advantage to this customized onsite construction of the nucleus besides enabling an exact fit between the closely rotating parts is, the dislodged proton particle could then establish orbit around the nucleus, just like planets orbiting around a star. If so, a balance or sphere of influence would be reached between the centrifugal force of the dark matter being flung away from the surface of the rotating nucleus material, and the pressurized dark matter fabric.

This balance suspends the micro-proton particle in an orbit around the nucleus, while energizing rotation. These revolving micro-proton particles or electrons, which are also in axial rotation, are held in different orbits based on a complicated relationship between the nucleus material and the companion electrons.

For instance, if the nucleus is flawlessly round and a perfect number of electrons occupy a single orbital layer, then the electrons would be held in an evenly spaced orbit and most likely have little or no ionic capacity.

Alternatively, an imbalanced nucleus configuration, containing neutrons and protons rotating on their axis, would produce uneven charges in the surrounding electron field resulting in the irregular spacing of the electrons. Additionally, within this varying electron field, each electron revolving around the nucleus would itself contain a sphere of influence, due to the electron's rotation on its axis.

Such energy spheres would result in a minimum spacing of the orbiting electrons in each layer, where if these electrons are held too close together, they will either combine or repulse one another depending upon the direction of their axial rotations. For example, if three nearby electrons are orbiting in the same layer too closely, their overlapping spheres of influence may act to repel one from the other.

In particular, the two electrons on either side with opposite axial rotation could work like a pump to force the middle electron out away from the nucleus, suspending the electron in the next orbital layer. Conversely, if two of the adjacent electrons exhibit like axial rotation, they may pull towards each other and combine, possibly resulting in the third smaller electron being thrown to the outer layer.

If an electron is forced out into the next orbital layer, its path will reach an equilibrium between the centrifugal force generated by the nucleus, the subsequent gravity produced from the spinning nucleus and, the sphere of influence from the inner electrons. In the first scenario, once this realignment occurs, the two remaining electrons left in the bottom layer will gain open space between them.

Nonetheless, their relative positions are locked in by the centrifugal and centripetal forces generated from the much larger rotating nucleus material, along with the like-vibrational pull between the electron and the proton, both made of the same material.

As for the second scenario, again the electrons will be locked in by the surrounding forces, but the outer orbital layer will end up further away from the nucleus, due to the wider sphere of influence generated by the initial combining of the inner two electrons.

This repulsion/ attraction reconfiguration process could utilize any number of combinations and most likely repeats itself throughout the electron orbital field during atom formation, until a balance either symmetrical or asymmetrical is established.

Such an arrangement adds stability to the atom's structural framework, permitting the electron's path to only deviate ever-slightly towards another approaching electron and then with each pass, is pulled back to the original position in a wriggle.

In practical terms, this energized open space surrounding the rotating and wriggling electrons can now act like a pump in reverse, by drawing or attracting micro-particles from outside of the atom, increasing the sphere of influence generated by the nucleus and determining the atom's overall ionic capacity. Given these two scenarios, the electron, being forced out and held in the outer orbital layer, will have a broader dimension of interaction with the inner electrons to vastly increase the atom's ionic capacity.

This added influence between the energetically overlapping electrons in the adjacent layers would imply a slightly variable electron mass, depending upon the proximity concerning the orbital position of each electron.

Primarily, one can visualize this energized space between the rotating, wriggling and revolving electrons, as either drawing inward towards the nucleus or pushing out away from the atom's center.

Therefore, the nature of the charge would be contingent on the electron field configuration, the spacing of the electrons and the direction of axial rotation between the coinciding electrons.

Such a malleable phenomenon, would predict the atom's ability to transform itself from attracting other atoms to repelling them, contingent upon the electron interplay. This possible duality, regarding the atom's charge, may solve the mystery behind what we have already identified as the elusive anti-matter force.

If protons and neutrons do actually start out with the same mass before combining to produce the electrons, when we factor in the mass of the electron, there is still some missing material. The possibility exists whereby this remaining material is pulverized into super micro-particles during the formation of the nucleus.

This as yet detected material, may be suspended in an orbit around the rotating electrons and would certainly add surface area to the atom, to create far greater ionic capacity. Holistically, once the atom is fully formed, the entire atom will now vibrate in a collective frequency cord, made up of the sum total of the atom's components.

This enables the like-matter process to reach the next level, which is the formulation of distinguishable mass into identifiable elements. Again, the dark matter fabric would work to juxtapose these like-vibrating atoms into close proximity where they could make ionic connections.

Hopefully, one can see how this inclusive theory unifies the macro workings of the universe with micro-particle fundamentals and ties them both to the energetic framework of the spirit realm, where equilibrium is sought through motion. Consequently, the vast quantity of energy residing in the spirit realm is the true engine driving the physical realm. Thus, requiring us to mathematically account for this phenomenon.

So too, we must consider the probability of this pervasive spiritual energy as coexisting and interacting with us in ways we are just now starting to understand. Bearing this in mind, a parting word worth mentioning is the idea of multiple dimensions within the spiritual realm, differentiated from one another by their energetic velocities.

Given light matter's propensity to travel at a widely variable rate, the soul would probably need to sustain a specific speed beyond its natural state in order to occupy a dimension in this non-physical setting. This may be a difficult feat as an independent entity possessing finite energy and, may relegate lone souls to reside in the less desirable state of transition when out of body.

Generating and sustaining such high orders of energy would probably require a synchronistic connection between multiple souls to supercharge the spirit energy's velocity. Ironically, connections between souls may only be established during life in the physical realm, where the slower moving energy is able link in an ionic way.

This potential paradox seemingly raises the bar for founding meaningful relationships during one's lifetime, in an effort to advance spiritually.

The Universal Frequency Theory, along with the corresponding equations were conceived and written by:

Kenneth W. Liskey Sr.

Science Points Worth Noting

The two-pole nature of this theory, lowers the energy requirements at the center. This would explain why the night sky is dark, because there is a low, slow burn and not a massive explosion. Also, the age of the universe must be reexamined, since we cannot interpolate back to a single event.

The Michelson-Morley experiment proved that no one-dimensional ether exists in empty space, but did not disprove the existence of the infinitely dimensional dark matter fabric, which could easily be responsible for transporting light through the empty space without detection.

The positive results for the laser-pulse clock experiments to determine whether space deviations affect time are actually proving that bent space, along with the varying gravity densities surrounding objects are affecting the path of the laser beam.

When both are calculated together, they will most likely account for the fractional differences in the time it takes for the laser beam to pulse from one point to the other. In any event, the laser beam, the equipment, the Earth and the observer are all expanding into the present moment during the experiment.

This does not negate the Theory of Relativity, for its application allows science to accurately predict the precise whereabouts of energy throughout space in the present moment.

Thermodynamics may have correctly assumed that all energy seeks equilibrium, but quite possibly has miscategorized heat dissipation as decay. In reality there is the action to heat and the equal and opposite reaction to cool. When taking a two-pole energy view, the dissipation of heat is merely returning the energy to the much larger closed loop dark matter energy grid for repurposing. This would be in keeping with the idea of energy's continuous ability to change form.

Antimatter forces seem far less mysterious when considering the cause and effect between rotation, centrifugal and centripetal forces, where the same mechanism can be made to either attract or repel, by manipulating this relationship. Such a comprehensive concept can be applied to both particle physics and astrophysics.

Under this theory, the hadron collider currently struggles to isolate an individual dark matter particle because the dark matter particles themselves are too tightly held within the vortex chains, making up each thread of the dark matter fabric. The irony being, these collective dark matter threads are what actually enables the hadron particles to reach such high speeds during the experiment.

A Demonstration in My Youth

When I was a child my father would take a large yellow plastic bucket, fill it halfway with water, and then with his arm straight down by his side, he would begin to whirl the bucket in a circular motion. Amazingly, when the pail reached the highest point of rotation above my dad's head the bucket would be upside down. Yet, the water never spilled out of the container. Rather, the centrifugal force compressed the water molecules tightly against the interior of the pail, making my dad a hero, for he could defy the laws of nature with one arm.

As a young man I was always troubled by such results, for science encourages us to believe the Earth's rotation generates gravity, which pulls everything towards the Earth. Nonetheless, here was this clear example of how rotation creates centrifugal force. Even more puzzling was that the Earth's gravity is generated by a large body of mass rotating at a high rate of speed, and I could not understand how my father could overcome such gigantic gravitational forces, single-handedly.

These problems were perplexing, until I considered the role of dark matter forces in this demonstration. For the rotation of mass not only flings dark matter out on a tangent away from planets, but this energy also pushes away from the pivot point of the rotating bucket and presses against the water molecules via centrifugal force.

However, when dark matter energy expansion, which is under constant pressure from the center of the universe, attempts to fill the void via gravity, it cannot overcome the centrifugal force held inside the swinging bucket. In essence, the rotating pail tethered to my dad's arm, contained not only the water, but also the centrifugally energized dark matter and it was this passive compression that prevented gravity from exerting its ground ward forces on the water.

Since we cannot rotate this small body of water without containment, it remains unclear how far out the water would expand from the center axis, before gravity would actually overcome this centrifugal force. However, it is clear when the water molecules are centrifugally compressed within the bucket, the earthward dark matter force is unable to fill the space between the water and the pail's inner surface.

Even so, the swinging bucket does not interrupt the gravity forces surrounding the bucket swinger. Otherwise, my father would have been flung away from the Earth on a tangent, as well. This demonstration proves we can mechanically manipulate the relationship between rotation, centrifugal force and gravity. In addition, this experiment may further our understanding of the mechanics inside of an atom. The diagram below has been modified slightly for illustrative purposes, so the man is now holding a rope which is tethered to the cutaway of a bucket containing the centrifugally energized water.

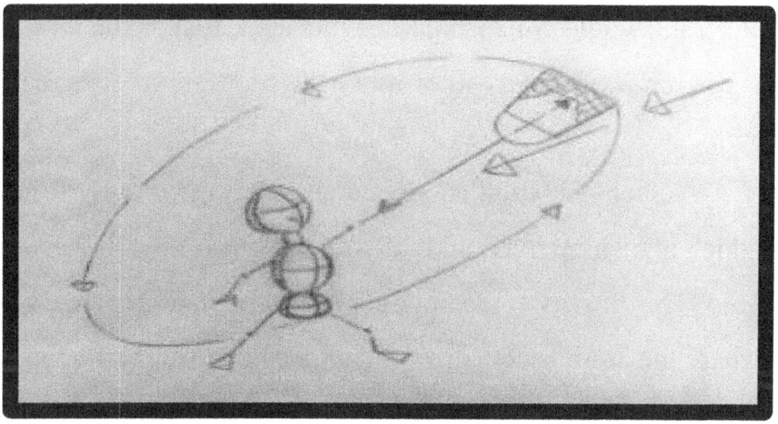

This man's rotating shoulder is analogous to the spinning components within a nucleus, where the reacting dark matter pushes the bucket away from the center, tightening the rope. The water contained in the bucket is like an electron, and is trapped in rotation. Fundamentally, the rope and bucket combination represent the nucleus's gravity coupled with the like-matter attraction between the proton and the electron, which locks the electron in an orbital layer. As such, the atom reaches a perfect balance between the centrifugal force outward, the centripetal force inward and the like-matter attraction to achieve the same effect.

Interestingly, when the man gets tired and the rotation of the bucket slows down, he gets wet because there is no longer enough outward force to hold the water in orbit. The same would be true for an atom, where the slowing electrons would collapse into the nucleus. Just as the man needs renewed energy to continue the process, so does the atom.

Such perpetual energy comes from the gravity force, which is rushing in behind the rotating bucket to fill the void. This gravity force not only holds the man in place, but where the atom is concerned, it abrades against the fleeing centrifugal force to spark energy into the process, keeping the electrons in orbit and the atom in continuous motion.

In all likelihood, dark matter particles operate and interact on a similar principal. Nonetheless, I was lucky enough to witness the awesome power of dark matter energy, which enabled my father to defeat the mighty Earth, if only for a moment.

The Mechanics of a Rotating Donut

There has been much written about disc-like flying spacecraft and we can use the analogy of a rotating donut, to describe the possibility of such a phenomenon. According to the Universal Frequency Theory, if a donut is free-rotating in space fast enough, the attached dark matter will fling out from the center on a tangent, in the direction of the rotation. So that, via centrifugal force, there will be a centrifugally charged band around the surface perimeter of the donut, (see diagram on the next page).

Additionally, the hole in the center of the donut is in rotation, but like the water in the yellow pail demonstration, the expanding dark matter is contained within the donut hole's perimeter. To relieve this centrifugal pressure, separate vortexes forms both above and below the donut hole. These vortices establish perpendicular ionic flow, which circulates through the center of the donut hole and then extends in a continuous loop, wrapping around the donut's outer wall, before returning back through the center.

These perpendicular dark matter energy bands flow tangentially around not only the donut wall, but they also engulf and trap the centrifugally charged dark matter band already established around the donut, creating a raised buffer zone surrounding the entire surface of the donut.

This ionically charged, elevated zone acts like a force field, preventing further dark matter interaction with the donut surface. However, the universe's energy fabric, would still interact with the outer edge of this ionically charged buffer zone. Hereafter, cause and effect takes over, where the expanding dark matter, pressurized from the center of the universe attaches itself to the charged buffer zone, and is then flung out on a tangent, forming a second centrifugally charged layer.

This second band would contain varying densities of dark matter, with less amounts along the leading edge of the rotating donut and progressively higher densities of dark matter, as measured around the donut's circumference, away from the leading edge. The reason being is, the corresponding gravity reaction to the donut rotation will lag.

Hence, a density imbalance is occurring within the space adjacent to the buffer zone, which can be exploited for space travel, permitting a vessel to maneuver into less dense space with extreme precision. This concept is similar to the volute sea shell idea, where the whirling water around the spiraling segments spreads out the water molecules and reduces resistance.

The overall end result from this rotating donut is that a vortex will be created through the donut's center, which will generate a vacuum at the point of suction, and produce an area of densified dark matter at the exit point. In particular, this equilibrium imbalance between suction and exit point will be instantly rectified by the densely congregated exiting dark matter, which will act to lift the donut into the vacated void or energy vacuum. In effect, we are describing the ultimate propulsion system.

Such an arrangement, where the expanding energy forced out from the rotating donut's perimeter is captured by a second perpendicular flow, while simultaneously generating an implosion through the center, is known as Torus flow. Just like other energy vortexes, the Universal Frequency Theory would predict that the raised centrifugal band along the donut's perimeter, abrading against the centripetal action through the donut's center, could squeeze the dark matter and coax a trickle release of the light matter energy stored within.

If this released light matter energy arc can be used to facilitate the donut's rotation, then external energy would only be required to initiate the donut rotation, along with periodic jump starts to maintain the rotational speed, depending on the efficiency of the donut. In this way, a spacecraft could travel great distances, with relatively low amounts of onboard energy. Furthermore, the Torus flow around the spacecraft, could be manipulated to reduce the density of dark matter in the direction of the travel, which would significantly lower the corresponding ionic resistance.

Lastly, what we have described is the bending of space around the donut. Accordingly, the rotating donut is actually generating a mini-portal around itself, and this folding of space, but not time, may be what places us within the range of alien beings from distant planets.

Energy Flow Around a Donut

A) Captured centrifugal bands

B) Perpendicular bands

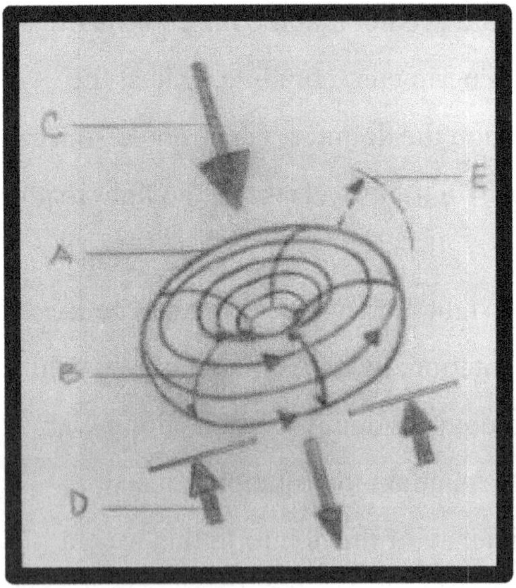

C) Vortex energy vacuum

D) Densified energy forces

E) Leading Edge of the second centrifugal band

Torus Energy Configuration

Cross Section View of a Dark Matter Energy Bundle

A bundle of dark matter energy threads is nearly impossible to encapsulate, but this image at least makes clear that each compressed thread is tightly stacked against the next one. What is not represented here is the Fibonacci spiraling of this entire bundle of threads, as it methodically winding its way out to the edge of the universe. Nor can one see the adjacent bundles which are interwoven with this thread to form the multi-directional dark matter fabric. Lastly, from this distant view, the tiny end-to-end vortices of each thread are invisible and appear in this image simply as individual spiraling lines.

Diagram of Dark Matter End-to-End Vortices

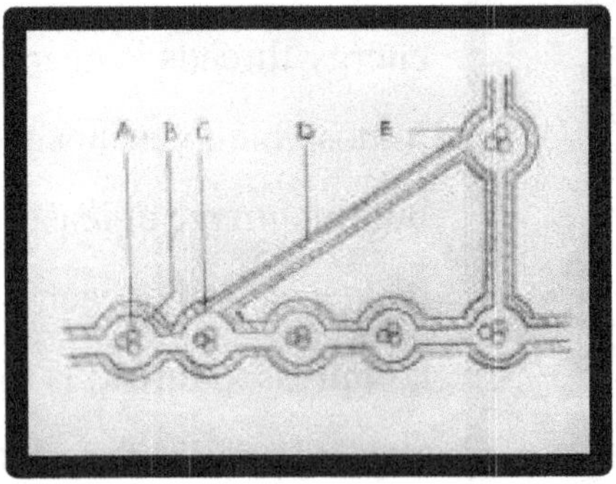

A) Dark matter particle groupings may behave similarly to the nucleus of an atom, whereby their axial rotations collectively create a sphere of influence.

B) Dark matter energy flowing around the particle groupings and, then arcing across to the next in-line group of particles as a vortex chain.

C) Light matter energy revolving around the dark matter energy on its way back towards the universe's center, flowing as a counter vortex thread.

D) A newly forming thread occurring between two existing threads, whereas the dark matter energy arc and, the coinciding light matter energy can cross the divide in any direction to generate new threads into the expanding space.

E) An existing end-to-end vortex thread running perpendicular to the bottom in-line thread. ***Note:*** The bottom vortex thread is expanding from left to right away from the center of the universe, as evidenced by the greater distance between the particle groupings. In effect, the particle groupings attract the energy, but cannot hold onto this fast flowing and highly pressurized current. Still, the particle rotation is what organizes the thread vortexes.

This paper was conceived and written by: Kenneth W. Liskey Sr.

The Copyright Registration Number for this theory is: TXu 2-282-950

Effective Date of Registration: September 28, 2020

Made in the USA
Monee, IL
08 November 2026

67994716R00042